YOUR KNOWLEDGE HAS VALUE

- We will publish your bachelor's and
  master's thesis, essays and papers

- Your own eBook and book -
  sold worldwide in all relevant shops

- Earn money with each sale

Upload your text at www.GRIN.com
and publish for free

Anupam Das, Ie Mei Bhattacharyya

Insights of Current Developments in Optics-Based-Biosensors for Analysis of Environmental Contamination and Pollution

GRIN Publishing

Bibliographic information published by the German National Library:

The German National Library lists this publication in the National Bibliography; detailed bibliographic data are available on the Internet at http://dnb.dnb.de .

Imprint:

Copyright © 2015 GRIN Verlag GmbH
Print and binding: Books on Demand GmbH, Norderstedt Germany
ISBN: 978-3-656-95700-3

This book at GRIN:

http://www.grin.com/en/e-book/299025/insights-of-current-developments-in-optics-based-biosensors-for-analysis

Insights of Current Developments in Optics-Based-Biosensors for analysis of Environmental Contamination and Pollution

Anupam Das

National Institute of Technology, Agartala

Department of Electrical Engineering

Ie Mei Bhattacharyya

National Institute of Technology, Agartala

Department of Electrical Engineering

Preface- As the process of industrialisation evolves, more and more pollutants add to the environment and put the health of the living world at stake. So the necessity to design user-friendly, advanced systems to detect and control these contaminants with high accuracy, sensitivity and stability has gained prime importance. Several conventional instruments are still in use but they have many drawbacks. Optical biosensors can be a suitable alternative to overcome the limitations of the conventional techniques as they are capable of real-time, high frequency monitoring of the pollutants with high accuracy, sensitivity, selectivity and precision. This article discusses about the various types of bio receptors which react to the analytes and how different techniques can be employed for their accurate detection by analysing the interaction of the bio receptors with the contaminants. This article also puts to light the remarkable developments in the field of optical biosensors and bio-recognition elements and their applications in pollution control.

Contents

1. Introduction..4

2. Bio-Recognition Elements used for Bio sensing purpose.....................................5

   2.1 The Enzymes as Bio-Recognition Element 5

   2.2 The Antibodies as Bio-Recognition Element 6

   2.4. The Aptamers as Bio-Recognition Element 7

   2.5. The Cells as Bio-Recognition Element 8

3. Use of Nanomaterials in Bio sensing purpose 8

   3.1. Use of Quantum Dots in Bio sensing purpose...8

   3.2. Use of Gold Nanoparticles in Bio sensing purpose......................................9

   3.3. Use of Graphene and Graphene Oxide in Bio sensing purpose 10

4. OPTICS-Based Biosenesors used for the purpose of environmental analysis.....................15

   4.1. Evanescent Wave Fiber Optic Biosensors..15

   4.2. SPR Biosensors in Environmental pollution analysis 17

   4.3. Nano-Structured Optical Biosensors in Environmental Pollution analysis...................18

5. New Emerging Optical Biosensors for Environmental Surveillance 18

   5.1. Optical ring resonator based biosensors 18

   5.2. Photonic crystal biosensors 18

6. Optical Biosensors for the purpose of surveillance of Pollution in the Environment............19

7. Conclusion 21

Reference...22

1. Introduction

In the present era of industrial development, several pollutants continue to contaminate the environment which effects the health of the nature as a whole. To meet this growing need to monitor these pollutants, many advanced analytical devices are being developed to detect and control these harmful substances. To detect the various contaminants that pollute water, quantitative analysis of water samples is done by chromatographic and spectroscopic methods. Despite being highly accurate and sensitive, these methods require expensive and highly sophisticated instruments, skilled personnel for handling, operation and to carry out the complicated procedure for the preparation of the samples. These methods also fail to accomplish real-time, onsite and high frequency monitoring of the pollutants. To overcome these drawbacks, extensive research is being carried out to devise cost-effective and dynamic monitoring techniques for accurate detection of these pollutants. Biosensors, which is designed from the combination of biochemistry, biology, nanotechnology, physics and electronics, exhibits all the desired characteristics like accuracy, speed, stability, low cost, real-time remote monitoring capabilities and has helped to improvise immensely in the area of risk management approaches and environmental issues.

A biosensor is an analytical device, used for the detection of an analyte that combines a biological component with a physicochemical detector. The Optical biosensors explores light absorption, fluorescence, luminescence, Raman scattering, reflection, and refractive index which are good substitutes to conventional techniques (Figure 1). These sensors provide fast, sensitive, real-time, and high-frequency monitoring without any time- consuming sample concentration. Even Though optical biosensors have huge potential applications in the fields of environmental monitoring, food safety, drug development, biomedical research, and diagnosis, but unfortunately their use in the areas of environmental pollution control is still in the infant stages [1-3]. Huge progress has been achieved in the research and development of optical biosensors, and a lot of research papers and outstanding reviews were published in the literature. This review mainly is centered on recent development and advancements in optical biosensors, providing examples of relevance, applications and the analytical performance in environmental pollution monitoring. For bio sensing bio-recognition molecules plays a vital part and are highlighted first. The important advances in optical biosensor will then be discussed and then new developments in the field of optical biosensors for pollution control will be reviewed.

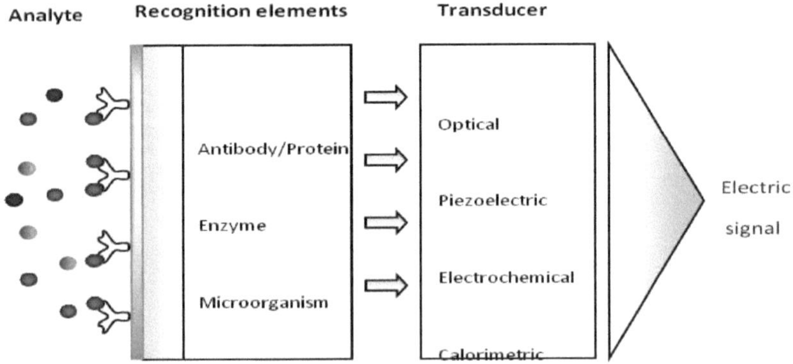

Figure 1. Schematic diagram of an optical biosensor.

2. Bio-Recognition Elements used for Bio sensing purpose

The important and key property of a biosensor is the construction of the bio-recognition element so that it could interaction with the targets molecules.

2.1 The Enzymes as Bio-Recognition Element

Enzymes are highly selective catalysts, greatly accelerating both the rate and specificity of metabolic chemical reactions, from the digestion of food to the synthesis of DNA. Optical biosensors based on enzymes have been rigorously in research during the last few decades due to the important practical based needs of industry for the purpose of environmental control and monitoring . Immobilization of enzymes on solid substrates is very vital because of the immobilization method do have the ability to increase the lifetime and sensitivity of the biosensors in practice. The optical transducers used in case of enzyme-based biosensors can be considered as the heart for the development of a compact, self-reliant device which could be used for environmental monitoring purpose. Cholinesterase (ChE) enzymes that can be inhibited by many toxic chemicals like organophosphates, heavy metals, and toxins. As a result of which, ChE biosensors are the main area interest in the context of global monitoring of toxicity. Taking in consideration that different pollutants inhibit enzyme activity in different ways, multi-analyte detection can also be done by using enzyme sensors. Taking an example, pesticides and heavy metal ions can be detected at a time in a single sample solution with inhibition of butyrylcholine esterase by pesticides and urease by heavy metals ions [4].

Enzyme based biosensor for the measurement of toluene in an aqueous solution was developed and characterized. In this case Toluene ortho-monooxygenase was being used as element of bio recognition, and an oxygen-sensitive ruthenium-based phosphorescent dye played the role of a transducer. Determination of Toluene for the purpose was based upon the enzyme-catalyzed consumption of oxygen that have the ability to change the phosphorescence intensity of the oxygen-sensitive probe [5-7]. Even though the enzyme based biosensor do have the ability to detect toluene in wastewater within a limit of detection (LOD) of 3.1 µM and a linear signal ranging up to 110 µM, but the response time is very long (1.2 h), and the activity goes on decreasing with each measurement. Huang *et al.* designed a fiber optic based biosensor to the determine adrenaline based on immobilization of laccase catalysis. The nanoparticles contained in laccase and the luminescent oxygen- sensing membrane were being deposited at the tip of the optical fiber. With the consumption of oxygen the enzyme laccase catalyzes the oxidation of adrenaline. The biosensor mentioned can detect adrenaline with in a range of 15 nM to 2 µM concentrations with a response time of 40 s. Immobilized enzyme is stable in nature. Enzyme-based optical biosensors provide new ways to perform remote, in-line determinations with high speed and accuracy for environmental pollution control and early warning. Although great improvement has been made in the reliability, sensitivity, selectivity and speed of response of these enzyme-based biosensors, certain limitations still exist. First, there are limited number of substrates that can be acted upon by specific enzymes; second, limited interaction between the environmental pollutants and specific enzymes; third, the enzymes fail to specifically differentiate among compounds of similar classes [8-9].

2.2 The Antibodies as Bio-Recognition Element

Use of particular interactions within antigen and antibody, the immunosensors have been considered as the novel-standard technique in environmental monitoring and clinical diagnostics. The highly certain interaction between two binding sites of an antibody with a specific target can be detected by means of a transducer (e.g.electronic,electrical,optical). As a result of which, the immunosensor provides a high repeatability, making it enable to recognize particular environmental pollutants [14-21]. Non-immunogenic environmental contaminants or pollutants having lower molecular weight (<1.2 kDa), which are called haptens, more or less becomes immune upon conjugation to carrier proteins . The Antibodies going against haptens, for example pesticides, persistent organic pollutants (POPs), and endocrine disrupting chemicals (EDCs), can be prepared by means of synthesizing the immunogens from the covalent binding of haptens into a carrier protein and then immunizing them with in animals [22].

The product related to the chemical binding of hapten to the carrier protein determines the specificity and quality of antibody which is very much important for immunoassay and it is called antigen. For the purpose of detecting the microcystin-LR (MC-LR), which is more frequent and mostly toxic hepatotoxin, the corresponding complete antigen (MC-LR-BSA) can be easily synthesized by introduction of a primary amino group in the seventh N- methyldehydroalanine residue of MC-LR. Then the resltant product aminoethyl-MC-LR can be coupled to bovine serum albumin (BSA) with glutaraldehyde. A monoclonal antibody (Clone MC8C10) against MC-LR can also be produced by means of immunizing with MC-LR-BSA [23-27]. Pollutants present in the environment with low molecular weight (molecular weight <1.2 kDa), and are much difficult to directly immobilize onto the biorecognition element sensing surface, results in antibodies that can be immobilized into preparation of the sensing surface for the immunosensors .

But, controlling the number, orientation, and the position of the antibodies in relevance to the sensor surface is a difficult task . Inadvertent disturbance on the binding site can occur if the antibody is in conjugate with the active surface of the sensor, which causes in the inevitable loss relating to the activity of the antibody. The most important point is , the use of strong acid in case of the regeneration process that reduces the capability of recognition of the immobilized antibodies after the reuse of the sensor surface, which ultimately affects the immunosensor interms of stability and reliability. The process of regeneration can be performed not more than 15 times, and the activity of the antibody tends to decrease in each cycle, resulting in inaccurate detection. To get a stable reusable sensor conjugation of the hapten-carrier-protein as bio- recognition element can be immobilized into the surface of the immunosensor . For instance, a reusable immune surface can be formed with the help of the covalent attachment of MC-LR-OVA to a self-assembled monolayer generated onto the fiber optic sensor with a heterobifunctional reagent. Process of regeneration on the surface of the sensor helps the performance of more than 110 cycles of assay devoid of any significant loss of reactivity which is less than 3% [26].

2.4. The Aptamers as Bio-Recognition Element

Aptamers are single-stranded DNA or RNA (ssDNA or ssRNA) molecules that can bind to pre-selected targets including proteins and peptides with high affinity and specificity. These molecules can assume a variety of shapes due to their propensity to form helices and single-stranded loops, explaining their versatility in binding to diverse targets. They are used as sensors, and therapeutic tools, and to regulate cellular processes, as well as to guide drugs to their specific cellular targets [29-30]. Contrary to the actual genetic material, their specificity and characteristics are not directly determined by their primary sequence, but instead by their tertiary structure. Interaction between the aptamer and the target is inclusive of the following, structure compatibility, stacking of aromatic rings, electrostatic and van der Waals interactions, hydrogen bonding, or a combination of all these effects. Because of the unique character of the aptamers they can be treated as useful alternative to a ntibodies as sensing elements [31]. Synthesis of Aptamers can easily be done chemically, and it does not require any complicated and expensive purification steps, resulting in elimination of the batch-to-batch variation found while using antibodies. Moreover, aptamers can undergo further modification through chemical synthesis for the purpose of enhancing the stability, affinity, and specificity of the molecules, also aptamers are more stable, and more resistant to denaturation and degradation than antibodies [32] . DNA/RNA aptamers are mainly designed for POPs, EDCs, organophosphorus pesticides biotoxins, and pathogenic microorganisms. Aptamers have become increasingly important bioassay materials for the purpose of environmental detection aptamers can play a vital role as bioassay material . According to a research literature reusable evanescent wave aptamer-based biosensor was developed for rapid, sensitive and highly selective detection of 17-β-estradiol, a natural endocrine based disrupting compound (EDC) having very high estrogenic activity. The β-Estradiol 6-(O-carboxymethyl)oxime-BSA was covalently immobilized into the surface of a optical fiber sensor. The dose-response curve of 17-β- estradiol was then developed with in a detection limit of 2.1 nM [33-39].

High specificity and selectivity of the optical biosensor were being exhibited by examining its response to a number of interfering endocrine-disrupting compounds. The Potential interference of real environmental sample matrices was examined using spiked samples in several wastewater effluents. The particular system described above can be used for on-site and real-time monitoring of 17-β-estradiol in wastewater treatment effluents or water bodies. There are various DNA aptamer fluorescence-based bio sensors which have been designed for the purpose of detection of Hg2+, Pb2+, and other trace of pollutants [46-49]. Kim *et al.* designed a DNA aptamer for detection of arsenic that can bind to arsenate [(As(V)] and arsenite [As(III)] with a dissociation constant of 4 and 8 nM, respectively . The use of the aptamer, a resonance scattering (RS)-based biosensor for the ultrasensitive detection of As(III) in aqueous solution via aggregation of gold nanoparticles (AuNPs) by the special interactions between arsenic-binding aptamer, target and cationic surfactant was being developed. The results shows that the variations of absorbance and RS intensity were exponentially related to the concentration of As(III) in the range from 1 to 1450 ppb, with the detection limit of 0.5 ppb for colorimetric assay and 0.78 ppb for RS assay [38].

There was another report on aptamer-based fluorescent biosensor for the highly selective and sensitive detection of Pb2+ and Hg2+ using a G-rich ssDNAs , which was labeled with the donor FAM at one end and the quencher DABCYL at the other end. The mentioned aptamer is in random-coil structure that gets converted into a G-quartet structure and a hairpin-like structure with the binding of Pb2+ and Hg2+, resulting moving of the fluorophore closer to the quencher and resulting in the decrease of the intensity of the fluorescence. The limits of detection of Pb2+ and Hg2+ ions are in the range of 0.2 nM and 4.0 nM, respectively. Though a variety of aptamers are under test for the detection of contaminants in real water samples [21].

2.5. The Cells as Bio-Recognition Element

The Whole cells are the best indicators of toxic components. Various microbial-based optical biosensors have been designed to detect toxicity and contaminants by monitoring the bio luminesce of light production or fluorescence . Olaniran et al. designed a whole- cell based bacterial biosensors for the fast and effective monitoring of heavy metals and inorganic contaminant in water . By the Use of Escherichia coli, the optical biosensors were found to be highly sensitive towards toxicity of wastewater effluents. The Bioluminescence increases with the increase in concentration of heavy metals and inorganic contaminants in water with a correlation coefficient of (r2) as high as 0.875 and 0.921, respectively. These bacterial biosensors can achieve the fast, sensitive and cost effective detection of contaminants in water [22-26].

Another research reported that an integrated fluorescence-based sensor for pH and oxygen [54], in which bacterial respiration was observed with the decrease in the oxygen partial pressure of the closed system and also with the decrease in pH value. Then the detection of inhibitory effect of toxic metal ions on the cellular operation of E. coli and Pseudomonas putida was achived. Amaro et al. reported a whole-cell biosensor for the detection of heavy metals based on metallothionein promoters from Tetrahymena thermophila . Two gene constructed with the help of Tetrahymena thermophila MTT1 and MTT5 metallothionein promoters linked with the eukaryotic luciferase gene, was being considerd as a reporter. The particular biosensor appeared to be the most sensitive eukaryotic metal biosensor among all other reported cell biosensors.The use of bioluminescent caused the bacteria immobilization in an alginate matrix on the bottom of the wells in a 87-well microplate, Eltzov et al. designed a fiber-optic biosensor for monitoring the toxicity of air . The Bioluminescence was being suppressed when the biosensor was exposed to toxic elements present in air Chloroform and that could be detected by this method with a LOD of 5.8 ppb [18].

3. Use of Nanomaterials in Bio sensing purpose

Nanomaterials exhibit unique size-tunable and shape-dependent physicochemical properties and have numerous possible applications in biosensors. The integration of nanomaterials and functional biological molecules (e.g., antibodies, nucleic acids, peptides) opens a new era in the optical biosensor field.

3.1. Use of Quantum Dots in Bio sensing purpose

Quantum dots (QDs), light-emitting semiconductor nanorystals, have been increasing used as biomolecular detection tools because of their unique optical properties, which conferred advantages over traditional fluorophores such as organic dyes. QDs have found applications ranging frombioanalytical assays, to live cell imaging, fixed cell and tissue labeling, and biosensors. The narrow, size-tuned, and symmetric emission spectra of QDs have made them excellent donors for fluorescence resonance energy transfer (FRET) sensors. Moreover, the overlap between the emission spectra of the donor and acceptor is reduced, and the cross-talk in such FRET pairs is circumvented . The broad excitation spectra of QDs facilitate excitation at a single wavelength far removed (>100 nm) from strategy employing water-soluble long lifetime fluorescence quantum dots and GNPs was used to detect trace Hg2+ ions in aqueous solution. The sensing system exhibits the detection limits their respective emissions, allowing QDs to be used in multiplex assays with single excitation sources. Using covalent or non-covalent linking approaches, the surface modification of QDs with antibodies, aptamers, and peptides are the most developed and widespread detection bioprobes. The long-term photostability, superior brightness, and good chemical stability of QDs enable them to greatly improve bioassay sensitivities and limits [61-63].However, controlling the number of antibodies (or aptamers) per QD as well as their orientation and position relative to the QD is difficult. Given the possibility of the inadvertent disruption of the binding site when QD conjugates with the antibody, the activity loss of the antibody is inevitable [60,64]. Additionally, antibodies usually need to be cryopreserved, but QDs cannot be frozen, thus making the storage of the QD-antibody a major obstacle to its practical application A carrier-protein-hapten-coupled QD nanobioprobe protocol has been developed to perform rapid and sensitive detection of small targets in environmental samples [31-34] .

The determination of 2,4-D in aqueous media was performed by grafting haptens-BSA conjugate on QDs and using the resulting material as a nano-bioprobe for 2,4-D biosensing. Samples containing different concentrations of 2,4-D were mixed with a given concentration of QD immunoprobe and fluorescence-labeled antibody, after which they were competitively detected by the all-fiber microfluidic biosensing platform. A higher concentration of 2,4-D resulted in less fluorescence-labeled anti-2,4-D antibody bound to the QD immunoprobe surface and consequently, lower fluorescence signal . The quantification of 2,4-D over concentration ranges from 0.5 nM to 3 M with a LOD of 0.5 nM. The method combined the merits of specific and stable binding interactions between environmental pollutants and its specific antibody, as well asthe excellent photophysical properties of QDs. The proposed immunosensor had the following unique advantages: first, QD-BSA-haptens conjugates used as recognition elements prevent compromise among the binding properties of the immobilized biomolecules (e.g., antibodies and enzymes); second, the binding sites of QD-BSA-haptens avoided steric hindrance and retained their high activity for their antibody; third, the structure of the QD-BSA-haptens conjugate was more stable in complex environmental samples than typical biorecognition molecules (e.g., antibody and enzymes). The FRET efficiency was higher because of the more abundant acceptor dyes bound to one QD surface, which conferred the QD-FRET assay with high sensitivity. This will provide a universal approach using a QD-bioconjugate as a nano-bioprobe to construct practical FRET-based immunoassay of various small molecules and other applications [32-33].

3.2. Use of Gold Nanoparticles in Bio sensing purpose

Gold nanoparticles (GNPs) with controlled geometrical, optical, and surface chemical properties have great potential applications in environmental and medical detection. GNPs can be easily modified with biomolecules. GNP-based optical biosensors commonly utilize fluorescence quenching through FRET or a visible color change attributed to the aggregation of AuNPs of appropriate sizes . GNP-based optical sensors have been used to detect environmental pollutants including heavy metals, toxins, and other pollutants . Heavy metal contaminations have greatly attracted public attention worldwide because of their serious negative health effects. Liu et al. used quaternary ammonium-functionalized GNPs to devise a colorimetric sensor for Hg2+ detection with the abstraction of GNP stabilizing thiols by Hg2+ inducing aggregation. An AuNP-rhodamine 6G-based fluorescent sensor was used to detect Hg2+, which had a LOD of 0.012 ppb . A T-Hg2+-T structure was used to develop a detection method of aqueous Hg2+ with a LOD of 50 nM , in which specific interaction of Hg2+ with thymine residues from two AuNPs induces the aggregation process and corresponding color change. Hg2+ and Ag+ could simultaneously be detected using FRET. However, this method was insufficiently sensitive for Hg2+ or Ag+ ion detection. Darbha et al. developed a AuNP-based sensor for the rapid, easy, and reliable detection of Hg2+ ions in aqueous solutions [71], which, through non-linear optical properties, had a LOD of 5 ppb (ng/mL). QDs have been utilized for FRET-based AuNPs assays for detection of environmental pollutants. An inhibition assay for identification of Pb2+ was developed based on the modulation in FRET efficiency between QDs and GNPs with a detection limit of 30 ppb of Pb2+ . The positively charged QDs form FRET donoracceptor assemblies with negatively charged GNPs by electrostatic interaction. The presence of Pb2+ aggregates AuNPs via an ion-templated chelation and inhibits the FRET Process [37-41]. A time-gated fluorescence resonance energy transfer (TGFRET) sensing of 0.49 nM in buffer and 0.87 nM in tap water samples .

3.3. Use of Graphene and Graphene Oxide in Bio sensing purpose

Fluorescent graphene-based materials have attracted great interest in the recent years due to their excellent chemical inertness, biocompatibility and low toxicity which makes them an appropriate choice for the detection of special targets. The range of graphene-based FRET biosensors' targets extends from DNA to ions, molecules and proteins through the integration of the functional biomolecules. Traditionally, the chemically derived graphene oxide (GO) has served as a predecessor for graphene but has drawn attention of many researchers for its own unique characteristics. The intrinsic and tunable fluorescence of GO can pioneer interesting and novel optical applications. A fluorescence sensor was reported for the detection of Ag(I) ions based on the target-induced conformational change of a silver-specific cytosine-rich oligonucleotide (SSO) and the interactions between the fluorogenic SSO probe and graphene oxide . Lee et al. used a platform based on chemiluminescence resonance energy transfer (CRET) between graphene nanosheets and chemiluminescent donors for homogeneous immunoassay of C-reactive protein (CRP). This 4. graphene-based CRET platform has a LOD of 1.6 ng/mL [41-43].

Liu et al. developed a homogeneous competitive fluorescence-based immunoassay for rapid and sensitive detection of microcystin-LR (MC-LR) based on the assembly of colloidal grapheme and MC-LR-DNA conjugates [78]. The MC-LR-DNA fluorescence probe was quickly adsorbed onto the graphene surface through the strong noncovalent - stacking interactions and can be effectively quenched through FRET. The competitive binding of anti-MC-LR antibody with MC-LR-DNA destroyed the graphene/MC-LR-DNA interaction, thus resulting in the restoration of fluorescence [43-44].

Table 1: A list of modifications to the SELEX process and their descriptions.

Method	Description	Reference
Atomic force microscopy (AFM)- SELEX	AFM-SELEX uses a dynamic atomic force microscopy tip to pick up and visualize aptamer-target complexes. This SELEX requires only one round of selection.	[21]
Automated SELEX	This SELEX uses automated systems for the procedure to reduce the time and labour required.	[23]
Blended SELEX	In this technique, a lead chemical compound is attached covalently or non-covalently to a nucleic acid library. Each nucleic acid conjugate in the starting library is a variant of the chemical compound moiety and allows up to 10^{15} variants of the small molecule to be screened for the most active of these composite assemblies.	[32]

Cell-SELEX	Cell-SELEX generates aptamers that can bind specifically to a cell of interest. Commonly, a cancer cell line is used as the target to generate aptamers that can differentiate that cell from other cancers or normal cells.	[33]
Capillary electrophoresis (CE)-SELEX	The separation of bound and nonbound oligonucleotides is performed using capillary electrophoresis.	[37]
Chimeric SELEX	Chimeric SELEX uses two or more different oligonucleotide libraries for production of chimeric aptamers with more than one wanted feature or function. Each of the parent libraries will be selected first to a distinct feature; the resulting aptamers are then fused together.	[37]
Conditional SELEX	This SELEX uses regulator molecules during the selection, thus, allowing aptamer binding to the target to be regulated.	[38]
Counter selection/ subtractive SELEX	This technique employs additional rounds of SELEX to remove sequences that bind to similar target structures.	[38]
Covalent/ Crosslinking SELEX	This process is used to select aptamers that contain reactive groups which are capable of covalent linking to a target protein.	[39]
Deconvolution SELEX	Deconvolution SELEX is used to generate aptamers for complex targets. Typically selection is performed on mixtures (or a cell). Once aptamers have been generated, a second part of SELEX involves discriminating which aptamers bind to which parts of the complex mixture.	[39]
Electrophoretic mobility shift assay (EMSA)-SELEX	The partitioning step of SELEX occurs through the use of electrophoretic mobility shift assay (EMSA) at every round.	[39]
Expression cassette SELEX	This is a special form of blended SELEX that involves transcription factors and optimizes aptamer activity for gene therapy applications.	[39]
Fluorescence-activated cell sorting (FACS) SELEX	This SELEX makes use of fluorescence-activated cell sorting to differentiate and separate aptamer-bound cells.	[40]
FluMag SELEX	Here the library is modified with fluorescein instead of radiolabels for quantification purposes. Additionally, the target is immobilized to magnetic beads instead of agarose.	[40]
Genomic SELEX	The SELEX library is constructed from an organism's genome and target proteins and metabolites from the same organism are used to elucidate meaningful interactions.	[41]

In vivo SELEX	In vivo SELEX uses transient transfection in an iterative procedure in cultured vertebrate cells to select for RNA-processing signals.	[41]
Indirect SELEX	The target used in the selection is not the aptamer binder; however, it becomes required for aptamer binding to the new target.	[41]
Mod-SELEX	Mod-SELEX uses a library of oligonucleotides with chemical substitutions that result in nuclease-resistant aptamers.	[41]
Multivalent aptamer isolation (MAI) SELEX	This process is used to generate aptamer pairs for a given target.	[42]
Microfluidics SELEX	This SELEX uses microfluidic technologies, creating an automatic, and miniature SELEX platform for fast aptamer screening.	[42]
Monolex	Monolex involves a single affinity chromatography step, followed by physical segmentation of the affinity material, to obtain the highest affinity aptamers.	[43]
Multiplexed massively parallel SELEX	This allows analysis of large numbers of transcription factors in parallel through the use of affinity-tagged proteins, bar-coded selection oligonucleotides, and multiplexed sequencing.	[44]
Multi-stage SELEX	Multistage SELEX is a modified version of chimeric selex. Here, the fused aptamer components then go through an additional selection with all the targets.	[45]
Negative selection	An additional step, performed typically at the beginning of selection, removes sequences that have an affinity for the selection matrix.	[45]
Next-generation SELEX	This SELEX uses designed oligonucleotide libraries that tile through a pre-mRNA sequence. The pool is then partitioned into bound and unbound fractions, which are quantified by a two-color microarray.	[46]
Non-SELEX (NCEEM)	This process involves repetitive steps of partitioning with no amplification steps.	[46]
Photo SELEX	Aptamers bearing photo-reactive groups that can photo cross-link to a target and/or photo activate a target molecule are used.	[46]
Primer-free SELEX	This SELEX involves removal of the primer-annealing sequences from the library prior to selection, preventing unwanted primer-based secondary structures.	[46]
Serial analysis of gene expression (SAGE) or	SAGE SELEX links oligomers from SELEX with longer DNA	[47]

high-throughput SELEX	molecules that can be efficiently sequenced.	
Spiegelmer technology	The aptamer selection is performed with the natural D-nucleic acids but on the opposite enantiomer of the chiral target molecule. After sequencing, the aptamers are synthesized as L-isomers for binding to the desired enantiomer of the target.	[52]
Slowoff-rate modified aptamers (SOMAmer)	The selection is performed with oligonucleotide libraries that are uniformly functionalized at the 5′-position resulting in high-quality aptamers.	[53]
Tailored SELEX	This is an integrated method to identify aptamers with only 10 fixed nucleotides through ligation and removal of primer binding sites within the SELEX process.	[53]
Target expressed on cell surface (TECS) SELEX	Recombinant proteins on the cell surface are used directly as the selection target.	[54]
Tissue-SELEX	This method is for generating aptamers capable of binding to tissue targets.	[56]
Toggle-SELEX	The selection is performed on different targets in alternating rounds.	[56]
Yeast Genetic SELEX	This method optimizes in vitro selected aptamers by creating a library of degenerate aptamers and performing a secondary selection in vivo using a yeast three (one)-hybrid system.	[56]

Table 2: A listing of DNA aptamers reported in the open literature* (up until July 2014) that have been confirmed to bind to small molecule targets. The dissociation constant, a measure of binding affinity, is included as well as the year of aptamer development.

Target	Binding affinity ()	Year	Reference
Reactive green 19	33 µM	1992	[12]
Adenosine monophosphate and adenosine triphosphate	6 µM	1995	[13]
L-arginine	2.5 mM	1995	[17]
L-argininamide	0.25 mM	1995	[17]

Anionic porphyrins	0.4–4.9 μM	1996	[17]
Sulforhodamine B	190 nM	1998	[19]
Cellobiose	600 nM	1998	[19]
7,8-dihydro-8-hydroxy-2'-deoxyguanosine	270 nM	1998	[20]
Cholic acid	5–67.5 μM	2000	[21]
Hematoporphyrin	1.6 μM	2000	[22]
L-tyrosinamide	4.5 μM	2001	[22]
Sialyllactose	4.9 μM	2004	[32]
Ethanolamine	6–19 nM	2005	[32]
(R)-thalidomide	1 μM	2007	[43]
Hoechst derivative 7e	878 nM	2007	[44]
17β-estradiol	0.13 μM	2007	[45]
Lys-Arg-Azobenzene-Arg	0.33 μM	2007	[46]
Tetracycline	64 nM	2008	[47]
L and D arginine	580–810 μM	2008	[46]
Daunomycin	10 nM	2008	[47]
Oxytetracycline	10 nM	2008	[45]
Ochratoxin A	200 nM	2008	[47]
Dopamine	700 nM	2009	[48]
8-hydroxy-2'-deoxyguanosine	100 nM	2009	[48]
Diclofenac	42.7–166.34 nM	2009	[48]
(S) and (R)-ibuprofen	1.5–5.2 μM	2010	[49]
Adenosine triphosphate	3.7 μM	2010	[52]
Fumonisin B_1 (FB$_1$)	100 nM	2010	[53]
Acetamiprid	4.98 μM	2011	[11]

Kanamycin	78.8 nM	2011	[9]
L-tryptophan	1.757 µM	2011	[51]
Bisphenol A	8.3 nM	2011	[53]
Ochratoxin A	96–293 nM	2011	[52]
Phenylphosphonic dichloride	>50 µM	2011	[58]
Organophosphorus pesticides (phorate, profenofos, isocarbophos and omethoate)	0.8–2.5 µM	2012	[59]
Polychlorinated biphenyls (PCB77)	4.02, 8.32 µM	2012	[59]
Polychlorinated biphenyls (PCB72 and PCB106)	60–100 nM	2012	[59]
Ampicillin	9.4–13.4 nM	2012	[59]

Only aptamer sequences that have experimentally determined values were included in this table.

4. OPTICS-Based Biosenesors used for the purpose of environmental analysis

4.1. Evanescent Wave Fiber Optic Biosensors

When a beam of light passes through an optical fibre and undergoes total internal reflection (TIR) at the core cladding interface, a thin electromagnetic wave (evanescent wave) is generated which decays exponentially with the distance from the interface with a typical penetration depth of several hundred nanometers .

$$E(z)= E0 \exp(- \delta / d p)$$

where δ is the distance from the interface, the penetration depth (dp) is given by:

$$dp = \lambda ex /2 \pi [(n2) \sin2 a- (n1)2]-1/2$$

where λex is the wavelength of the light, n1 the refractive index of the cladding region and n2 the refractive index of the core and α is the angle of incidence measured from the normal at the interface of the core and cladding. The generated evanescent wave excites fluorescence in the vicinity of the sensing surface e.g., in fluorescently labelled biomolecules bound to the optical sensor surface through affinity recognition interactions. The ability of the evanescent wave to discriminate between the bound and the unbound fluorescence complexes due to its short range, eliminates the normally required washing procedures. Moreover, biomolecular interactions can be well-studied and detected by evanescent field-based waveguides [44-47]. Evanescent wave fiber-optic immunosensors (EWFI) exhibit fast response, high sensitivity, low cost and are suitable for real-time on-site detection. They can be used to detect a wide variety of pollutants such as TNT, 2,4-D, atrazine, E. coli O157:H7, and

Staphylococcal enterotoxin B . Conventional EWFI consist of numerous optical components (e.g., chopper, off-axis parabolic reflector, and biconvex silica lens) which makes it heavy and expensive. Crucial optical alignment restricts its use as portable device. An evanescent wave all-fiber biosensor (EWAB) is developed which is simple in construction, compact and portable which is based on a single multi-fiber optical coupler for simultaneous detection of 2,4,-D and MC-LR (Figure 2).

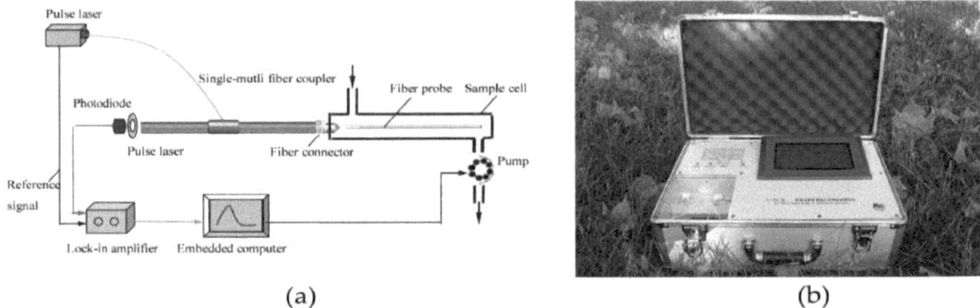

(a)

(b)

Figure 2. Diagram of the portable evanescent wave optical fiber biosensor (a) Diagram describing the Working principle of the portable optical fiber biosensor and (b) the platform.

The use of a single multi-fiber optical coupler for both the transmission of the excitation light and the reception and transmission of fluorescence, reduced the number of components and eliminated the requirement of optical alignment. It thus resulted in a significant signal enhancement. Tube-etching method was used to produce combined tapered fiber probes which were modified by covalent attachment of the MC-LR- OVA (recognition element) to a self-assembled monolayer formed onto the probe. This probe is highly resistive to non-specific binding of proteins and can be reused more than 150 times with a LOD of 0.03 µg/L and a LOD of 0.07 µg/L for MC-LR and 2,4-D, respectively [47-49] .

The EWAB based on QDs and TIRF was used for ultrasensitive DNA detection which

resulted in an exceptional detection limit of 3.2 amol DNA. The ssDNA coated probe was covalently immobilized onto a self-assembled alkanethiol monolayer of fiber optic probe through a streptavidin-biotinylated ssDNA strategy. A 30-mer ssDNA, the segments of the uidA gene of E. coli., was detected. The probe can be reused for more than 30 assay cycles. Based on the proposed theory, a quantitative measurement of DNA binding kinetics was performed with high accuracy, indicating an association rate of 1.38×10^6 M1s1 and a dissociation rate of 4.67×10^3 s1. Thus it is evident from the above mentioned facts that optical biosensing platform provides a simple, cost-effective, fast and robust solution for clinical diagnosis, pathology and genetics [48-51].

A highly toxic and ubiquitous pollutant, mercury ions ($Hg2+$), require fast and highly sensitive on-site detection techniques in food products and environment. In order to protect the environment and health, a portable, low-cost and fast heavy metal analysis system for initial on-site/in situ screening of heavy metal-contaminated sites is prioritised. An evanescent wave all-fiber optical biosensor based on structure-switching DNA for rapid on- site/in situ detection of heavy metal ions has been reported . A DNA probe that can hybridize a fluorescently labelled complementary DNA containing a T-T mismatch structure was covalently immobilized onto a fiber optic sensor. If the sample contains mercury ions, part of the fluorescence-labelled DNAs bind with Hg2+ to form T-Hg2+-T complexes through the folding of the DNA probe segments into a hairpin structure and dehybridisation from a fiber optic sensor, resulting in a decrease in fluorescence signal. The total analysis time for a single sample, including measurement and surface regeneration, was less than 10 min with a detection limit of 1.2 nM. This sensing strategy may be an alternative for the analysis and assessment of the transport and fate of environmental pollutants. In our previous paper , based on a direct structure-competitive detection mode, an evanescent wave DNA-based biosensor was also used for rapid and sensitive detection of Hg2+ with a detection limit of 2.1 nM. The sensor surface can be regenerated over 100 times with no significant deterioration of performance [50-53].

A proof-of-concept development of optic fiber-based immunoarray biosensor was demonstrated for the detection of multiple small analytes . Through the immobilization of two kinds of hapten conjugates (MC-LR-OVA and NB-OVA) onto the same fiber optic probe, MC-LR and TNT could be detected simultaneously and specifically within an analysis time of approximately 10 min. The LODs for MC-LR and TNT were 0.04 and 0.09 mg/L, respectively. Good regeneration performance, binding properties, and robustness of the sensor surface of the proposed immunoarray biosensor ensure the cost-effective and accurate measurement of small analytes.

4.2. SPR Biosensors in Environmental pollution analysis

Surface Plasmon Resonance (SPR) is a surface-sensitive optical technique that is associated with the evanescent electromagnetic field generated on the surface of a thin metal film when excited by an incident light under total internal reflection conditions . Due to the fact the evanscent field diminishes exponentially with increasing distance of penetration from the interface, SPR promotes monitoring of only surface-confined molecular interactions occurring on the transducer surface. Most of the SPR instruments use a Kretchmann configuration working at attenuated total reflectance (ATR) for excitation of surface plasmons, which can detect a small refractive index change at the metal/analyte interface, and the information of the molecular interactions can be obtained by measuring the optical intensity (or phase/polarization) of light reflected from the optical instrument. SPR biosensors allow real-time detection of minute changes in the refractive index when biorecognition molecules (e.g., antibodies) immobilized on a transducer surface bind with their biospecific targets (e.g., analytes) in solution. Since their introduction in the early 1990s, SPR biosensors have seen wide applications including clinical diagnosis, drug discovery, food analysis, environmental monitoring. In general, a SPR biosensor is comprised of several important components: a light source, a detector, a transduction surface (e.g., gold-film), a prism, biorecognition molecule (e.g., antibody/antigen, DNA and aptamer) and a flow system [49-52].

The use of SPR to detect environmental contaminants, including atrazine, Dichloro- Diphenyl-Trichloroethane (DDT), 2,3,7,8-tetrachlorodibenzo-p-dioxin, carbaryl, 2,4-D, benzo[a]pyrene (BaP), biphenyl derivatives, and trinitrotoluene (TNT), has recently gained considerable interest [10,86-89]. An SPR immunosensor for BaP, a carcinogenic endocrine disrupting chemical, was reported to have a LOD of 10 ppt . A portable SPR-based immunosensor was developed for the analysis of carbaryl in natural water samples . Based on a binding inhibition immunoassay format, this immunosensor has a LOD of 1.38 g/L. The sensor surface covalently modified by the analyte derivative allows the reuse for more than 220 regeneration cycles. The immunoassay performance of the biosensor was validated with respect to conventional high-performance liquid chromatography-mass spectrometry, and the correlation between methods was in good agreement (r2 > 0.998) for real water samples. Kim et al. fabricated the sensing surface of the SPR immunosensor simply by covalent amide binding of 2,4-D-

BSA conjugate on the Au-thiolate self-assembly. A LOD of 0.1 ppb 2,4-D is established with a response time of only 4 min.

4.3. Nano-Structured Optical Biosensors in Environmental Pollution analysis

Progress in nanotechnology, microelectronics and microfluidics could facilitate development of miniaturized, rapid, ultrasensitive and inexpensive nano-structured optical biosensing platforms for rapid toxicity screening and multianalyte testing. These devices are likely to become more compact, robust, smaller and adaptable for in-field and continuous field-based environmental monitoring monitoring. A fiber-optic nanosensor was designed with taper optical fibers, onto which biorecognition molecules (e.g., antibody, peptides, and nucleic acids) was immobilized. This sensor can probe individual chemical species in a living cell . In situ measurements of the carcinogen BaP in a single cell could be achieved by a fiber-optic immuno-nanosensor , and the quantitative detection ranges from 1.56×101 M to 1.56×108 M [42-45].

A mesoporous silica nanosensor, which responds selectively to $Fe2+$ (pH = 8) and $Cu2+$ (pH = 12) with a distinguishable colour change perceivable by the naked eye with a detection limit of approximately 50 ppb was synthesized by the co-condensation method . A whispering gallery mode (WGM) nanosensor consists of an optical resonator and a circular cavity. A tapered optical fiber placed next to the cavity was used to introduce light evanescent coupling. Armani et al. developed a WGM nanobiosensor using a micro-toroid cavity with a Q greater than . In this study, single-molecule detection sensitivity for antibody-antigen binding was demonstrated. This nanobiosensor could perform real time single-molecule detection and exhibited a dynamic range from 5 aM to 1 M.

5. New Emerging Optical Biosensors for Environmental Surveillance

5.1. Optical ring resonator based biosensors

Optical ring resonator is an emerging sensing technology, in which at least one is a closed loop coupled some sort of light input and output . In a ring resonator, the light propagates in the form of whispering gallery modes (WGMs) or circulating waveguide modes. When light of the resonant wavelength transports through the loop from input waveguide, it builds up in intensity over multiple round-trips due to constructive interference and is output to the output bus waveguide which serves as a detector waveguide. The WGM spectral position is related to the refractive index (RI) through the resonant condition: = 2rneff / m, where r is the   ring outer radius, neff the effective RI experienced by the WGM, and m is an integer. neff changes when the RI near the ring resonator surface is modified due to the capture of target molecules on the surface, which in turn leads to a shift in the WGM spectral position. Thus, by directly or indirectly detecting the WGM spectralshift, the quantitative detection of targets will be achieved [39].

5.2. Photonic crystal biosensors

Biosensors are enabled superior levels of sensitivity resulting in precise detection limits. Photonic crystal biosensors are very small and are possible through coupling the incident and reflected/transmitted light to optical fibers and analyzing them in remote locations.

Photonic crystal fibres have wavelength-scale morphological microstructures that run along the entire fiber length by corralling it within a periodic array of microscopic air holes.Overcoming the limitations of conventional fiber optics, photonic crystal fibers are proving 18 State of the Art in Biosensors - Environmental and Medical Applications to have a multitude of important technological and scientific applications including biosensors. Due to their well-defined physical properties such as reflectance/ transmittance, photonic crystal biosensors are enabled superior levels of sensitivity resulting in precise detection limits. Photonic crystal biosensors are very small and are possible through coupling the incident and reflected/transmitted light to optical fibers and analyzing them in remote locations.

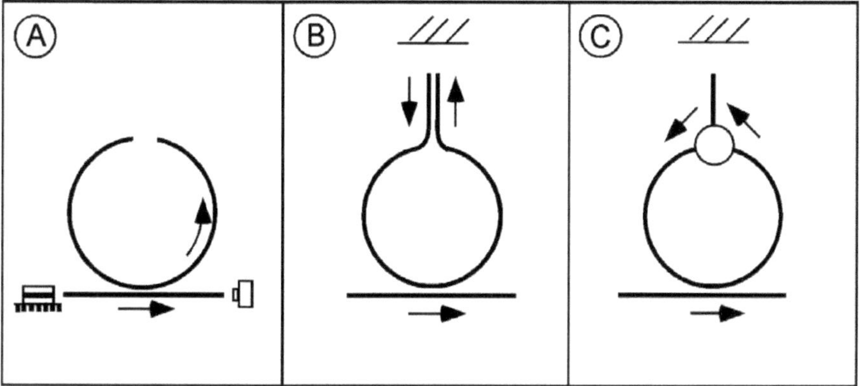

Figure 3. Various ring resonator biosensors

To design a photonic crystal biosensor, some portion of the resonant electric field must be in contact with liquid media that contains the analyte, providing a surface on which biorecognition molecules may be adsorbed. Label-free photonic crystal biosensors generally detect shifts in resonant wavelength or coupling angle caused by the interaction between the target molecule and the evanescent wave. The narrow spectral linewidth achieved by using high Q factor passive optical resonators enables sensor systems to resolve smaller wavelength shifts associated with the detection of analytes at low concentration, such as environmental pollutants. Photonic crystals biosensors have been applied for sensing the pH and ionic strength of solutions, metal ions and trace orgnic pollutants [54-56].

6. Optical Biosensors for the purpose of surveillance of Pollution in the Environment

Due to rapid growth of industries, the generation of industrial wastes has been intensified to a large extent over the last few years. This has led to an increased amount of pollutants in surface and ground waters. Stringent pollution control and regulatory acts set up by various environmental pollution control authorities has resulted in development of a non-costly general pollution contol and regulatory network system. An Early Warning System (EWS) represents an integrated system which monitors,analyzes,interprets and communicates data monitoring using the concept of real-time detection in a continuous mode. It also involves triggering of an alarm when any detection of pollutant is encountered in the water which in turn, accomplishes the identification of the events of low probability/ high-impact contamination in a realistic time period that takes into account the safety of public health and hygiene [57].Quick response and precautionary warning, automated sampling, automatic detection with a viable range of sensitivity and minimum amount of false positives/false negatives are the desirable features of an integrated EWS ideally. So far, optical biosensors are found to be the much more effective in multitarget sensing as compared to other sensors. On-site real- time monitoring of a contamination in continuous manner is also known to be done by these sensors much more effectively. Accidental spillage or any such type of pollution entity contaminating water requires mapping for which optical biosensors are being used. In this regard these biosensors are needed to be integrated into several EWSs. Monitoring of contamination in both drinking and surface water by online mode are nowadays efficiently done by luminescent bacteria based biosensors. Several strains of bacteria have been reported till date which play an important role in detectory part of testing contamination based on a wide range of parameters of toxicity levels namely heavy metal content, effluents containing organic pollutants,e.t.c. This has led to the development of a bioluminescent bacterial biosensor consisting of multi-channeled network which helps in detecting of heavy metals present in water online.This technique led to the online detection of 0.5 M CdCl2 and 5 M As2O3 from an incoming stream when four bacteria of

bioluminescent type (E. coli DH1 pBzntlux, pBarslux, pBcoplux, and E. coli XL1 pBfiluxCDABE) were used collectively. Cross-detecting of a single or many heavy metals was done simultaneously by this biosensor online which also did the sample's overall toxicity measurement . Changes in fluorescence spectral patterns and kinetics of the algae are the key factors that help in continuous determination of toxic products in water. This forms the basis of working of the bbe Algae toximeter.The test is performed by te addition of automatically and independently cultured standardized algae to the sample of water followed by the analysis of active chlorophyll concentration. An alarm is triggered in case of damage of the algae by the action of herbicides that are responsible for the reduction in activity or by the evolution of oxygen. Monitored water sample by this system leads to higher time-scale resolution which reflects the system's high sensitivity for the recognition of several herbicides and their by-products. TOX control system used freshly cultivated Vibrio fischeri bacteria as a biological sensor. Inhibition percentage is computed by the measurement of luminescence before and after exposition. Light loss from the suspended mixture of luminescent bacteria to a greater extent is observed with the increased toxicity level of the sample. Combination of the precision of instrumentation system and toxicity detection and testing of the organism makes this system more attractive and advantageous.

Development of an automated water analyzer computer-supported system (AWACSS) has been initiated which does not require any pretreatment procedure for the sample and help fast measurement of organic pollutants in trace amounts. A network of measurement and control stations is being used by this early-warning system which comprises of the four main components namely: the AWACSS instrument with fluid control and optical transducer chip, the HTC PAL auto-sampler for preparing sample, the personal computer at the sampling site, and the server with database and Web site . Selectivity of this system helps in analyzing trace quantities of toxic materials in complex matrices also. Fast multi- target detecting of estrone, propanil, isoproturon, atrazine, bisphenol A, sulphonamides, and progesterone had been done by applying this system. Most of the targets had the detection limits in the range of few nanogram or sub-nanogram per litre Internet based networking between the measurement and control stations, global management, trend analysis, and early-warning applications are known to be the basis of working of a Web-based AWASS system.Recently, development of an automated online optical biosensing system (AOBS) with a futuristic approach was initiated which helped in fast detection and warning of microcystin-LR (MC-LR). Premixing of samples of varying concentrations of MC-LR with a specific concentration of fluorescence-labeled anti-MC-LR-MAb is done. Fluorescence- labeled anti-MC-LR-MAb is highly specific to MC-LR enabling its binding. The whole thing is performed with the usage of an indirect competitive mode which is followed by the pumping of the sample mixture onto the surface of biochip modified by MC-LR-ovalbumin. Higher concentration of MC-LR resulted in lower fluorescence signaling owing to less fluorescence-labeled antibody. The quantified range of MC-LR is observed as 0.2 g/L to 4 g/L and the LOD is quantified to be 0.0 9 g /L [58-59]. The biochip consists of six sensing parts which enables the AOBS for determining six different pollutants in matrices existing in environment simultaneously when fluorescence-labeled antibodies are being utilized by other analyte conjugates. This is done after modification of each sensing point by the analyte conjugates. Scheduled analysis meets the requirement of maintaining a permissible safety limit of sources of drinking which accomplishes the objective of AOBS.

7. Conclusion

Research and developments in the field of optical biosensor engineering has improved our ability to characterize and quantify contaminants and pollutants present in the environment, which without doubt offers huge bonus for the purpose of environmental pollution control. The Optical biosensors do have many important advantages for environmental pollution control and monitoring applications : (1) optical biosensor helps in providing a very fast, simple, sensitive, and selective assay method; (2) ability to provide prolong-period, automatic high-frequency measurements of environmental pollutants and contaminents; (3) biosensing arrays enable the development of nano, robust, and adaptable optical biosensors for fast screening of toxicity and multiple-analyte testing at a same time of the pollutants and contaminants present in the environment ;(4) The incorporation of nanoelectronics and microfluidics into optical biosensors will help in miniaturizing optical biorecognition elements; (6) wireless-communication engineering facilitates the development and designing of environmental sensor based networks; (7) the long-term, high-frequency online pollutant monitoring ability of the optical biosensors provides a new height into the development and migration mechanism and the future of environmental pollutants incorporated with the physicochemical parameters. Though there are various other challenges which are to be delt with, for production of more reliable biosensors, but undoubtly optical biosensors will strive to provide the most productive paths for environmental pollution control and early warning.

Reference

1. Bellan, L.M.; Wu, D.; Langer, R.S. Current trends in nanobiosensor technology. *WIREs Nanomed. Nanobiotech.* 2011, *3*, 229-246.

2. Thevenot, D.R.; Toth, K.; Durst R.A.; Wilson, G.S. Electrochemical biosensors: Recommended definitions and classification. *Biosens. Bioelectron.* 2001, *16*, 121-131.

3. Borisov, S.M.; Wolfbeis, O.S. Optical biosensors. Chem. Rev. 2008, 108, 423-461.

4. Dorst, B.V.; Mehta, J.; Bekaertb, K.; Rouah-Martin, E.; Coen, W.D.; Dubruelc, P.; Blusta,R.; Robbens, J. Recent advances in recognition elements of food and environmental biosensors: A review. Biosen. Bioelectron. 2010, 26, 1178-1194.

5. Ligler, F.S. Perspective on optical biosensors and integrated sensor systems. Anal. Chem. 2009, 81, 519-526.

6. Kim, S.J.; Gobi, K.V.; Tanaka, H.; Shoyama, Y.; Miura, N. A simple and versatile self-assembled monolayer based surface plasmon resonance immunosensor for highly sensitive detection of 2,4-D from natural water resources. Sens. Actuators B 2008, 130, 281-289.

7. Kawaguchi, T.; Shankaran, D.R.; Kim, S.J.; Matsumoto, K.; Toko, K.; Miura, N. Surface plasmon resonance immunosensor using Au nanoparticle for detection ofTNT. Sens. Actuators B 2008, 133, 467-472.

8. Vo-Dinh, T. Nanosensing at the single cell level. Spectrochim. Acta Part B 2008, 63, 95-103.

9. Vo-Dinh, T.; Griffin, G.D.; Alarie, J.P.; Cullum, B.; Sumpter, B.; Noid, D. Development of nanosensors and bioprobes. J. Nanopart. Res. 2000, 2, 17-27.

10. Moorty, M.S.; Cho, H.J.; Yu, E.J.; Jung, Y.S.; Ha, C.S. A modified mesoporous silica optical nanosensor for selective monitoring of multiple analytes in water. Chem. Commun. 2013, 49, 8758-8760.

11. Armani, A.M.; Kulkarni, R.P.; Fraser, S.E.; Flagan, R.C.; Vahala, K.J. Label-free, single-molecule detection with optical microcavities. Science 2007, 317, 783-787.

12. Hasan, J.; Goldbloom-Helzner, D.; Ichida, A.; Rouse, T.; Gibson, M. Technologies and Techniques for Early Warning Systems to Monitor and Evaluate Drinking Water Quality: A state-of- the-art Review; EPA/600/R-05/156; U.S. Environmental Protection Agency: Washington, DC, USA, 21 September 2005.

13. Woutersen, M.; Belkin, S.; Brouwer, B.; van Wezel, A.P.; Heringa, M.B. Are luminescent bacteria suitable for online detection and monitoring of toxic compounds in drinking water and its sources? Anal. Bioanal. Chem. 2011, 400, 915-929.

14. Charrier, T.; Chapeau, C.; Bendria, L.; Picart, P.; Daniel P.; Thouand, G. A multi-channel bioluminescent bacterial biosensor for the on-line detection of metals and toxicity. Part II: Technical development and proof of concept of the biosensor. Anal. Bioanal. Chem. 2011, 400, 1061-1070.

15. Zurita, J.L.; Jos, A.; Cameán, A.M.; Salguero, M.; López-Artíguez, M.; Repetto, G. Ecotoxicological evaluation of sodium fluoroacetate on aquatic organisms and investigation of the effects on two fish cell lines. Chemosphere 2007, 67, 1-12.

16. Tschmelak, J.; Proll, G.; Riedt, J.; Kaiser, J.; Kraemmer, P.; Wilkinson, J.S. Automated Water Analyser Computer Supported System (AWACSS) Part I: Project objectives, basic technology, immunoassay development, software design and networking.Biosens. Bioelectron. 2005, 20, 1499-1508.

17. Shi, H.; Song, B.; Long, F.; Zhou, X.; He, M.; Lv, Q.; Yang, H. Automated online optical biosensing system for continuous real-time determination of microcystin-LR with high sensitivity and specificity: Early warning for cyanotoxin risk in drinking water sources. Environ. Sci. Technol. 2013, 47, 4434–4441.

18. Jang, A.; Zou, Z.; Lee, K.K.; Ahn, C.H.; Bishop, P.L. State-of-the-art lab chip sensors for environmental water monitoring. Meas. Sci. Technol. 2011, 22, 1-18.

19. Lee, J.S.; Joung, H.; Kim, M.; Park, C.B. Graphene-basedchemiluminescence resonance energy transfer for homogeneous immunoassay. ACS Nano 2012, 6, 2978- 2983.

20. Liu, M.; Zhao, H.; Chen, S.; Yu, H.; Quan, X. Colloidal graphene as a transducer in homogeneous fluorescence-based immunosensor for rapid and sensitive analysis of microcystin-LR. Environ. Sci. Tech. 2012, 46, 12567-12574.

21. Jung, J.H.; Cheon, D.S.; Liu, F.; Lee, K.B.; Seo, T.S. A graphene oxide based immuno-biosensor for pathogen detection. Angew. Chem. Int. Ed. 2010, 49, 5708-5711.

22. Andrade, J.D.; Vanwagenen, R.A.; Gregonis, D.E. Remote fiber-optic biosensors based on evanescent- excited fluoro-immunoassay: Concept and progress. IEEE Trans. Electron Devices 1985, 32, 1175-1179.

23. Golden, J.P.; Saaski, E.W.; Shriver-Lake, L.C.; Anderson, G.P.; Ligle, F.S. Portable multichannel fiber optic biosensor for field detection. Opt. Eng. 1997, 36, 1008- 1013.

24. Long, F.; He, M.; Zhu, A.; Song, B.; Sheng, J.; Shi, H. Compact quantitative optic fiber-based immunoarray biosensor for rapid detection of small analytes. Biosens. Bioelectron. 2010, 26, 16- 22.

25. Cooper, M.A. Optical biosensors in drug discovery. Nat. Rev. 2002, 1, 515-528.

26. Miura, N.; Sasaki, M.; Gobi, K.V.; Kataoka, C.; Shoyama, Y. Highly sensitive and selective surface plasmon resonance sensor for detection of sub-ppb levels of benzo[a]pyrene by indirect competitive immunoreaction method. Biosens. Bioelectron. 2003, 18, 953-959.

27. Mauriz, E.; Calle, A.; Abad, A.; Montoya, A.; Hildebrandt, A.; Barcelo, D.; Lechuga, L.M. Determination of carbaryl in natural water samples by a surface plasmon resonance flow-through immunosensor. Biosens. Bioelectron. 2006, 21, 2129-2136.

28. Darbha, G.K.; Singh, A.K.; Rai, U.S.; Yu, E.; Yu, H.T.; Ray, P.C. Selective detection of mercury(II) ion using nonlinear optical properties of gold nanoparticles. J. Am. Chem. Soc. 2008, 130, 8038-8043.

29. Wang, X.; Guo, X.Q. Ultrasensitive Pb2+ detection based on fluorescence resonance energy transfer (FRET) between quantum dots and gold nanoparticles. Analyst 2009, 134, 1348- 1354.

30. Huang, D.; Niu, C.; Ruan, M.; Wang, X.; Zeng, G.; Deng, C. Highly sensitive strategy for Hg2+ detection in environmental water samples using long lifetime fluorescence quantum dots and gold nanoparticles. Environ. Sci. Tech. 2013, 47, 4392-4398.

31. Rao, C.N.R.; Sood, A.K.; Subrahmanyam, K.S.; Govindaraj, A. Graphene: The new two-dimensional nanomaterial. Angew. Chem. Int. Ed. 2009, 48, 7752-7777.

32. Liu, Y.; Dong, X.; Chen, P. Biological and chemical sensors based on graphene materials. Chem. Soc. Rev. 2012, 41, 2283-230.

33. Wen, Y.Q.; Xing, F.F.; He, S.J.; Song, S.P.; Wang, L.H.; Long, Y.T.; Li, D.; Fan, C.H. A graphene-based fluorescent nanoprobe for silver(I) ions detection by using graphene oxide and a silver-specific oligonucleotide. Chem. Commun. 2010, 46, 2596-2598.

34. Hollenstein, M.; Hipolito, C.; Lam, C.; Dietrich, D.; Perrin, D.M. A highly selective DNAzyme sensor for mercuric ions. Angew. Chem. Int. Ed. 2008, 47, 4346-4350.

35. Xiang, Y.; Tong, A.; Lu, Y. A basic site-containing DNAzyme and aptamer for label-free fluorescent detection of Pb2+ and adenosine with high sensitivity, selectivity, and tunable Dynamic range. J. Am. Chem. Soc. 2009, 131, 15352-15357.

36. Li, T.; Wang, E.; Dong, S. Lead(II)-induced allosteric G-quadruplex DNAzyme as a colorimetric and chemiluminescence sensor for highly sensitive and selective Pb2+ detection. Anal. Chem. 2010, 82, 1515-1520.

37. Wang, X.D.; Wolfbeis, O.S. Fiber-optic chemical sensors and biosensors (2008-2012). Anal Chem. 2013, 85, 487-508.

38. Olaniran, A.O.; Hiralal, L.; Pillay, B. Whole-cell bacterial biosensors for rapid and effective monitoring of heavy metals and inorganic pollutants in wastewater. J. Environ. Monit. 2011, 13, 2914-2920.

39. Arain, S.; John, G.T.; Kranse, C.; Gerlach, J.; Wolfbeis, O.S.; Klimant, I. Characterization of microtiterplates with integrated optical sensors for oxygen and pH, and their applications to enzyme activity screening, respirometry, and toxicological assays. Sens. Actuators B 2006, 113, 639-648.

40. Amaro, F.; Turkewitz, A.P.; Martín-González, A.; Gutiérrez, J.C. Whole-cell biosensors for detection of heavy metal ions in environmental samples based on metallothionein promoters from Tetrahymena thermophila. Microb. Biotechnol. 2011, 4, 513-522.

41. Eltzov, E.; Pavluchkov, V.; Burstin, M.; Marks, R.S. Creation of a fiber optic based biosensor for air toxicity monitoring. Sens. Actuators B 2011, 155, 859-867.

42. Eltzov, E.; Marks, R.S.; Voost, S.; Wullings, B.A.; Heringa, M.B. Flow-through real time bacterial biosensor for toxic compounds in water. Sens. Actuators B 2009, 142, 11-18.

43. Cui, Y.; Wei, Q.; Park, H.; Lieber, C.M. Nanowire nanosensors for highly-sensitive selective and integrated detection of biological and chemical species. Science 2001, 293, 1289-1292.

44. Wang, L.; Ma, W.; Xu, L.; Chen, W.; Zhu, Y.; Xu, C.; Kotov, N.A. Nanoparticle-based environmental sensors. Mater. Sci. Eng. R 2010, 70, 265-274.

45. Medintz, I.L.; Clapp, A.R.; Mattoussi, H.; Goldman, E.R.; Fisher, B.; Mauro, J.M. Self-assembled nanoscale biosensors based on quantum dot FRET donors. Nat. Mat. 2003, 2, 630-638.

46. Wang, L.; Liu, X.; Zhang, Q.; Zhang, C.; Liu, Y.; Tu, K.; Tu, J. Selection of DNA aptamers that bind to four organophosphorus pesticides. Biotechnol. Lett. 2012, 34, 869-874.

47. Jo, M.; Ahn, J.Y.; Lee, J.; Lee, S.; Hong, S.W.; Yoo, J.W.; Kang, J.; Dua, P.; Lee, D.K.; Hong, S.; et al. Development of single-stranded DNA aptamers for specific bisphenol a detection. Oligonuleotides 2011, 21, 85-92.

48. 49. Kim, Y.S.; Jung, H.S.; Matsuura, T.; Lee, H.Y.; Kawai, T.; Gu, M.B. Electrochemical detection of 17β- estradiol using DNA aptamer immobilized gold electrode chip. Biosens. Bioelectron. 2007, 22, 2525-2531.

50. Peterson B. Moraes; Rodnei Bertazzoli; Ederio Dino Bidoia ,Electrochemical oxidation of wastewater containing aromatic amines using a flow electrolytic reactor, International Journal of Environment and Pollution, 2013 Vol.51, No.1/2, pp.1 - 14.

51. A. Ramesh; B.N. Nagendra Prakash; P.V. Sivapullaiaih, Identification of source of heavy metal contamination in a site - a case study, International Journal of Environment and Pollution, 2013 Vol.51, No.1/2, pp.91 - 105.

52. B. Fisher, J. Kukkonen, M. Schatzmann, Meteorology applied to urban air pollution problems: COST 715, Int. J. of Environment and Pollution, 2001 Vol.16, No.1/2/3/4/5/6, pp.560-570.

53. Medhat Ibrahim, Hanan El-Haes, Computational spectroscopic study of copper, cadmium, lead and zinc interactions in the environment, Int. J. of Environment and Pollution, 2005 Vol.23, No.4, pp.417 - 424.

54. Ahmet Altin, Sureyya Altin, Birol Elevli, Mustafa Degirmenci, Ali Kemal Ozdemir, The effect of BOD5 and flow-rate of activated sludge systems on investment and energy costs, Int. J. of Environment and Pollution, 2005 Vol.23, No.4, pp.425 - 437.

55. Selma Ates, Muge Kilic, Removal of phenol using mushrooms and immobilized polyphenol oxidase in a plug flow reactor, Int. J. of Environment and Pollution, 2005 Vol.23, No.4, pp.480 - 485.

56. Weimin Xie, Qunhui Wang, Hongzhi Ma, Hiroaki Ogawa, Phosphate removal from wastewater using aluminium oxide as adsorbent, Int. J. of Environment and Pollution, 2005 Vol.23, No.4, pp.486 - 491.

57. Bok Haeng Baek, Viney P. Aneja, Observation based analysis for the determination of equilibrium time constant between ammonia, acid gases, and fine particles, Int. J. of Environment and Pollution, 2005 Vol.23, No.3, pp.239 - 247.

58. Raffaele Molinari, GianLuca Scicchitano, Fabrizio Pirillo, Vittorio Loddo, Leonardo Palmisano, Preparation, characterisation and testing of photocatalytic polymeric membranes with entrapped or suspended TiO2, Int. J. of Environment and Pollution, 2005 Vol.23, No.2, pp.140 - 152.

59. Joshua Ifeanyichukwu Ume, Influence of solute dissociation constants on the optimum adsorption of weak electrolytes on activated carbons, Int. J. of Environment and Pollution, 2005 Vol.23, No.2, pp.215 - 222.

Figures:

Figure 1:   https://yildizuludag.files.wordpress.com/2010/06/biosensor.jpg
Figure 2:   http://www.intechopen.com/books/state-of-the-art-in-biosensors-environmental-and-medical-applications/recent-progress-in-optical-biosensors-for-environmental-applications
Figure 3:   http://www.intechopen.com/books/state-of-the-art-in-biosensors-environmental-and-medical-applications/recent-progress-in-optical-biosensors-for-environmental-applications